Game Theory

(The Everyday Guide)

How to Think Strategically, Make Good Decisions and Improve your Life

Table of Contents

Introduction

Game theory refers to the science of strategy and strategic decision-making. It works by logically and mathematically determining the actions that people should take in order to produce the best outcome from their decisions. If you want to succeed in life, then having enough knowledge about strategically making decisions can help you a lot.

In life, there is always a need to make decisions. If you do not know how to make decisions logically, then you and the people around you will be drastically affected. By learning the game theory, you will have an idea about how to go about things and think strategically. You will have better chances of beating the odds and succeeding in life.

Game theory also works in looking at the relationships between the participants of a specific model with an aim of predicting their optimal decisions. The theory is mainly applicable to actual games. It studies different types of games including tennis, child-rearing and takeovers. The games studied by the game theory share a common feature and that is interdependence.

What it means is that the output of every participant in the game is dependent upon the strategies implemented and the choices made by everyone. You can also use the game theory in analyzing decisions. This involves acquiring information that will be extremely useful in your attempt to formulate the soundest decisions and produce the best results out of them.

If you are interested to learn more about the game theory and how you can apply it not only in actual games, but also in your life, then this book is for you. It will talk about the different topics relevant to the theory including its basics and the way it can help you in effective decision-making.

This book will also introduce you to the concepts of bargaining, group decisions, cooperative and fair division, all of which are essential in fully understanding game theory and implementing it to your life. With the help of this book, you will become an expert in game theory and learn how you can use it to improve your relationships with other people.

You can also start making strategic decisions that will help you achieve success and form great relationships with the people in your office.

Chapter 1: The Basics of Game Theory

In essence, game theory refers to a vital process which involves modeling strategic interactions between more than two players in a specific situation wherein a set of outcomes and rules are in place. It is a popular theory in numerous disciplines, but it is a tool used mostly in the field of economics. Applying game theory in the field of economics serves as a valuable tool when there is a need to do effective fundamental analysis of various industries, strategic interactions between more than two companies and various sectors.

It is important to note that game theory does not refer to the study of the basics of winning a game like chess or creating role playing scenarios for different types of games and sports. In fact, this popular theory does not significantly relate to things that you commonly refer or consider to as games. Basically, it refers to the study of the way nations, people and firms identify strategies and techniques that they can use to deal with the competing strategies and techniques carried out by the other parties they are dealing with.

The parties, firms and nations involved in a game theory are known as players or agents. The theory also makes an assumption that the key players or agents formulate logical and rational decision all the time. However, some experts believe that this is a false assumption. The main reason is that what the majority of the society believes as irrational behavior like building up nuclear weapons is actually rational when taking into account the standards of game theory.

Note, however, that despite the ability of game theory analysis to produce counter-intuitive outcomes and results, it is still capable of yielding surprising insights regarding human nature. The theory delves deeper into the specific reason why a scenario happens.

An example is analyzing the real reason why the members of a particular society cooperate with each other. In this case, the game theory tries to figure out whether the cooperation is only designed to achieve material gain, or if there are more reasons aside from material gain.

The game theory is extremely useful when facing a situation involving at least two agents or players. The situation should also involve known quantifiable consequences or payouts. You can use the theory to analyze and figure out the most probable outcome from a particular scenario. To make it easier for you to understand game theory, check out the following basic concepts commonly used in this field of study:

Game - It refers to a set of scenarios or circumstances with a result or an outcome that fully depends upon the actions carried out by at least two players, agents or decision-makers.

Players/Agents - This refers to the parties involved in the game. They are the decision-makers who need to use logical, strategic and rational thinking to make the soundest decision and produce the best outcome.

Strategy - This refers to the plan of action carried out by the player or agent in the course of achieving the best outcome to a particular game or scenario.

Payoff - This refers to the reward or payout the player or agent can get after reaching a specific result or outcome. Payoff should be in a quantifiable form. For instance, utility or dollars.

Equilibrium - This refers to the point wherein all the players involved in a game were able to formulate a decision and reached their desired outcome.

Information set - This term refers to all the available information provided at specific points of the game. This is usually applicable when a sequential component is present in a game.

Now that you have an idea about the basic concepts used in game theory, it is time to delve deeper into this field. You will learn more about this valuable tool in the succeeding chapters.

Chapter 2: Basic Decision-making Game Theory

As mentioned in the earlier parts of this book, game theory plays a vital role in strategic and effective decision-making. You can actually find a number of game theory strategies that will help you formulate the best decision for a particular situation. Game theory is extremely effective in decision-making that companies are already using it to formulate high reward or high risk strategic decisions especially when they are facing highly competitive situations and markets.

A lot of those who are already using the theory for a long time work on leveraging its principles with the help of strategic games. The strategic games used in this case refer to well-defined mathematical situations encompassing a set of agents or players (usually refer to firms or individuals), set of techniques that the players can use, as well as a reward or payoff specification for every combined strategy.

Game theory works perfectly in strategic situations wherein modeling individual or competitive behaviors is a possibility. Examples of such situations are bargaining activities, product decisions, supply chain design, principal-agent decisions and auctions. Basically, it involves playing multiple strategic games as a means of modeling various competitors, potential strategies and payoffs.

The goal of playing the game is to produce a reliable and recommended set of strategic and sound decisions that will guide certain competitive behaviors to reach desirable outcomes. It also aims to analyze how the set of potential strategies can forecast different competitive results. It is possible to use various kinds of games based upon the presented strategic scenario, the amount of available data or information, timing limitations or constraints and number of players.

When planning to use game theory for decision-making, take note that it usually makes an assumption that players will take rational actions based on their individual interest. The theory also makes an assumption that all players intend to take strategic actions and take into consideration the competitive results or responses of each action that they take.

Another important thing to take note of is that game theory works more effectively if those who implement it fully understand the payoffs, both negative and positive, of every action that they make. Game theory is also applicable in various fields that require effective and strategic decision-making. Examples of such fields are economics, psychology, finance, business and political science.

With the aid of its principles, you can significantly reduce the risk of every action or step you take. One game theory strategy that you can use in this case is the Prisoner's Dilemma. It refers to a concept which works by exploring the decision-making strategies implemented by two players who acted on their best self-interest, causing the outcome to become worse than if they decided to cooperate with each other.

The typical scenario is established in a way that the two parties decide to protect their own selves even if this move does not produce the ideal outcome to the other party. Because each party followed a logical thought process, which involves helping only their own selves, both find themselves getting the worst output.

Chapter 3: Bargaining, Fair Division, Cooperative and Group Decisions

Four of the most popular concepts in game theory are bargaining, fair division, cooperative and group decisions. It is crucial to gain a full understanding of these concepts if making the most out of applying game theory strategies into your life is your goal.

Bargaining in Game Theory

Bargaining in game theory actually involves two players or agents. It involves dealing with a problem focused on figuring out how the two players can cooperate in case the most probable result from non-cooperation are Pareto-inefficient outcomes. In essence, this concept refers to an equilibrium selection problem. It is crucial to note that most games come with numerous equilibria, each one with different payoffs or reward for every player.

This forces the players involved in the bargaining problem to make negotiations regarding the equilibrium to target. The bargaining concept also has an underlying principle that the resulting outcome or solution should be similar to the solution recommended by an impartial arbitrator. Possible solutions to this concept can be categorized into two approaches.

The first one is the axiomatic approach, which satisfies the preferred properties of a proposed solution. The second one is the strategic approach, which involves modeling the bargaining process in full detail and as a sequential game.

Fair Division

Fair division is another concept that you have to understand fully if you want to take full advantage of the game theory. This concept is applicable when facing a problem that involves the need to divide a set of resources or goods between a number of people who are entitled to them. This means that every person involved needs to have his or her own fair or due share.

This kind of scenario happens in numerous real-world settings. Few examples are divorce settlements, airport traffic management, allocation of electronic spectrum and frequency and auctions. The core principle behind the fair division concept is that the players should perform the process of dividing the problematic resources or goods themselves. They can actually seek the aid of a mediator.

However, they should avoid seeking the aid of an arbiter for this, since the players are the only ones who know how valuable the goods or resources are.

Cooperative and Group Decisions

The cooperative concept of the game theory means that the game is composed of groups of players who will most likely enforce cooperative behavior. This results to the game turning into one wherein coalitions or groups of players compete, instead of the individual players. One specific example of this concept is the coordination game. It involves picking strategies or techniques based on the consensus decision-making procedure done by the players.

It is important to note that recreational games do not normally fall under the cooperative category. The main reason is that these games frequently lack mechanisms that allow the players of a coalition or group to enforce a cooperative behavior. However, you can see an abundance of the required mechanisms in actual or real-life scenarios. This makes the cooperative concept a vital element in game theory.

Chapter 4: Static Complete Information Games

Static complete information games are among the games that you will encounter as you work on mastering the game theory. This actually covers games with simple forms. The first step is that the players pick actions simultaneously. They can then expect to receive a reward or payoff depending on the combined actions they have just chosen. The core principle behind the static games of complete information is that the payoff function of every player, which refers to that which identifies the payoff of a player from his chosen combined actions, is already a common knowledge. This means that all the other players are already aware of it.

Note, however, that even if the players have the chance to pick their preferred strategies or techniques simultaneously, it is not crucial for all the parties involved to take action simultaneously, as well. It is also crucial to note that rational players are not inclined to play dominated techniques strictly, even if the payoff is already somewhat a common knowledge. The main reason is that there is no proof that the player who chose the combined actions can hold that combination for long.

When trying to understand static complete information games, note that most economists find the process of modeling static strategic interactions as games wherein players obtain continuous or infinite strategy spaces. One concrete example of this is the Bertrand competition. This involves numerous profit-maximizing companies that tend to compete for a spot in a heterogeneous or homogeneous market. They do so by selecting their prices simultaneously.

If any of the players decides to increase the price, then there is a great possibility for their income per unit to increase, as well. However, this will also lead to a sudden reduction in the number of products sold. This output will have a great impact on competitors.

Another example is the Cournot competition. This is where numerous companies that aim to maximize their profit compete in a market by selecting their output's scale simultaneously. A decision to increase the output will most likely result to increased sales. However, this move can also have a great impact to competitors since it has the tendency to lower market price.

Static games of complete information are also applicable in a political setting. This takes place when there is competition in the political scene. For instance, political parties whose aim is to maximize their votes tend to compete by picking their advertising campaigns or efforts simultaneously.

Their decision to raise the number of their advertising campaigns will lead to an increase in the amount that they will spend for it. However, this move is also favorable for the party who decided to increase his advertising campaign, and unfavorable for the other parties, because it boosts the chances of the former to succeed.

Chapter 5: Dynamic Complete Information Games

Dynamic complete information game is another of the most popular concepts in game theory. In several vital economic applications, it is necessary to think about playing the game for multiple time periods. This act can make the game dynamic. There are two reasons that can turn a game into a dynamic one.

The first one is that the player's interaction is naturally dynamic. This means that the players are capable of observing the actions taken by other players prior to deciding the optimal response that they should take.

The second reason that can turn a game into a dynamic one is if it is a one-off game, which is repeated numerous times. It also requires the players to observe the result of the past games prior to playing the newer ones.

One thing that differentiates dynamic complete information games from the other is that a player has already observed all the past moves prior to choosing his next move. The player is also fully aware of who has already moved what, and he can use this information to formulate a decision.

A dynamic game of complete and perfect information can also be represented by using the normal-form. This makes it necessary to use a Nash equilibria set that appears in its normal form. Finding the set of Nash equilibria in this particular dynamic game is possible by constructing the normal form of the game.

If a dynamic game is represented in extensive form, then note that you can predict what will most likely happen by translating the extensive form into an associated normal form. After the translation, you can then implement the Nash equilibria concept. Take note that dynamic complete information games normally have numerous Nash equilibria.

Another thing to take note of is that a lot of Nash equilibria found in dynamic games typically involve those players who choose the non-credible techniques. The core principle in dynamic complete information games is to ensure the credibility of the strategies chosen by the players. This makes it necessary to use a strong solution that will rule out any non-credible techniques or strategies.

One solution that you can use is the sub-game perfect Nash equilibrium. This solution works by specifically adhering to the sequential rationality principle. This means that equilibrium techniques should be able to specify the most ideal behavior from any point, whether reached or not, not only at the equilibrium path, but also in the future game.

Chapter 6: Static Incomplete Information Games

Numerous economically important scenarios involve a game, which starts with some of its players holding private details or information about anything that is relevant to making a decision. If you encounter this scenario, then note that it falls under games of incomplete information. However, you should avoid confusing incomplete information with imperfect information. Note that the latter is considered imperfect since the involved players did not conduct a proper observation of the actions taken by the others.

Despite the fact that the players are not fully aware of some of the private information of the opponent, they can still make some solid assumptions on what their opponents know. Some situations offer the chance of modeling the informational asymmetry just by understanding that every player is fully aware of his or her payoff function. The problem is that such player is not sure about the payoff functions of the other players of the game or his or her opponents.

To increase your chances of winning static incomplete information games, it is advisable to understand a player's type. Note that the player's type completely describes all information that he has that is not yet a common knowledge. All players are fully aware of their own type with full certainty. The beliefs that a player has about his or her opponents are also usually based upon a joint probability distribution that is already common knowledge over the other types of player.

You can also perceive the game as one which starts with an act related to nature. While a nature's act is most likely imperfectly observed, every player is still capable of observing the specific type bestowed upon him by nature. However, other players will be unable to observe directly the type bestowed by nature upon others.

In this case, effectively implementing the Bayesian equilibrium to any static incomplete information games that you encounter can help. It refers to a strategic profile which allows each player type to maximize his anticipated utility based upon his opponents' type-contingent techniques, as well as the probability distribution bestowed over each player type.

It is also crucial to note that numerous issues found in a static incomplete information game can make it a bit challenging to assess which among the players have different information. These issues include the actual number of players who are involved in the game, the most feasible actions or moves of each player, the outcome based upon the actions taken or chosen by players, and the preferences of each player regarding possible outcomes.

It is crucial to understand all these issues if you want to have an easier time knowing exactly what your opponents are hiding. In static incomplete information games, after observing and realizing that each player can simultaneously pick actions based on their own type, each one can start making a move. The game will end with payoffs based upon the type of player, as well as the specific action that he chooses.

Chapter 7: Dynamic Incomplete Information Games

Dynamic incomplete information games actually cover a lot of interesting economic models. A few examples are reputation, cheap talk and market signaling. Among the problems that you may encounter when dealing with these types of game are those related to sub-game perfection. Note that the sub-game perfection requirement does not seem to work very well when used in an extensive form game that has incomplete information.

One issue that you may encounter in this case is the failure of the sub-game perfection to rule out sub-optimal actions despite providing beliefs regarding uncertainty. The problem usually comes out when one side of the game does not come with any sub-games, except for the actual game. This makes it impossible for the sub-game perfection requirement to work. A potential solution for this problem is to make it a requirement for all players to make an optimal choice in every information set involved in the game.

Another problem that you may encounter in dynamic incomplete information games is that the sub-game perfection may permit actions or moves that are only possible if these come with beliefs that are categorized as "unreasonable". One of the best solutions for the problems that usually arise in dynamic incomplete information games is the PBE (Perfect Bayesian Equilibrium).

This concept implies that the player who has picked a move at every information set should have a belief regarding the node in the set that was already reached when playing the game. A probability distribution set over the information set's nodes represent this concept.

However, problems still arise when implementing this concept especially because some PBE examples permit equilibria that seem a bit unreasonable. Note that PBE does not pose limitations on certain beliefs applicable to situations that have zero probability. The good news is that game theorists were able to introduce numerous equilibrium refinements as a means of ruling out unreasonable PBE's. Such refinements work in limiting the kinds of beliefs held by all players in certain situations with zero probability.

It is also possible to eliminate unreasonable or implausible equilibrium by setting a requirement for all players to have beliefs then using such beliefs in testing and analyzing the sequential rationality of the strategies implemented by each player.

Another scenario that you will most likely encounter in dynamic incomplete information games is the signaling game. It implies that one player holds private information, which allows him to take action that can greatly influence the utility of every possible type he makes. Another player observes the actions of the first player, and base his own actions on what he observed. After each player took their own action, the payoffs are finally realized.

Solving signaling games or other similar games that fall under the dynamic incomplete information category is a bit hard and challenging. This holds especially true if you try to solve it in a formulaic way. You may also find it hard to find or discover all equilibria. This is only possible if you think deeply about the problem or issue presented to you.

It helps if you use all the things you know regarding the kind of equilibrium you are searching for. For instance, when dealing with a separating equilibrium, you have an idea about the exact beliefs that are relevant to it. This knowledge can help ensure that you will not be dealing with unprofitable or unreasonable deviations.

Chapter 8: Game Theory in Everyday Life

Game theory is a useful concept that you can apply in your daily life. It allows you to use strategies that make it possible for you to make rational decisions. Also, note that game theory is a concept wherein every decision that you make, as well as the decisions of other players in the game that you are playing, have their individual consequences. The decisions of others also have the tendency of helping you shape yours.

The good thing about game theory is that it has a number of strategies that you can use to improve your everyday life. There are also numerous situations in your life wherein you will find the theory useful. This chapter discusses some of these scenarios.

To Save Money

Everyone wants to save money in one way or another, and the game theory can help you with that. For example, you can apply the principles of this theory to buy a car at a more affordable price. You can expect the principles of the theory to work better in this case than simply visiting a car dealership and trying so hard to haggle for the price. The first step in this example is to find every car dealership that offers the model and brand you are currently looking for.

The next step is to call each of of the car dealership and say that you will purchase the car from a dealer who offers you the lowest price. If a particular dealer responds by telling you that their company does not make negotiations over the phone, then you can simply say that it is possible to negotiate over the phone since you know several deals that were closed that way. You can expect some dealers to go the extent of offering you a lower price than what the others offer, so as to lure you into buying from them.

With this example, it is safe to assume that game theory really plays a big role if you intend to use it to save money. Apply the same technique whenever you need to negotiate and want to save money as much as possible. You will be surprised on how this technique works in letting you enjoy huge savings in various situations.

Dealing with Heavy Traffic

If you are a someone with an extremely hectic schedule, then the heavy traffic is probably one of the things that annoy you. Questions like whether it is best to change lanes, when will the car that is in front of you will change routes or when the lane beside you will start to speed up will come across your mind. If you are impatient, then there is a great possibility that you will immediately switch lanes. This is a good move, but only if most of the drivers in your lane are not impatient like you.

However, if most of them are also the impatient type, which prompts them to move into the fast lane, then there is a great possibility that this will eventually turn into the slow lane. In this case, you may find that the situation is hopeless since you will realize that every action that you take will only work if the others won't do the same. The good news is that the game theory presents an effective solution that tends to work out for everyone who faces the same dilemma. This solution is to choose randomly when is the perfect time to switch lanes.

What you have to do to take full advantage of this solution is to check the other lane every few minutes. If you notice that it moves faster, then consider switching right away. However, if you noticed that the other lane is slowing down, then it is best to stay where you are. Continue doing the process every few minutes, until you are able to pick the best course of action for the situation you are in.

Experts believe that if all drivers implement this solution, then there is a great possibility for everyone to end up in an equilibrium. This is the particular state wherein both lanes move at similar speed. While reaching an equilibrium is not the best payoff for you, it is still the most reasonable and fair output.

Doing Household Chores (Example: when to wash the dishes)

Household chores are among the most dreaded activities of people. Some hate doing household chores like washing the dishes causing such tasks to pile up. If you are someone who hate doing household chores, then questions like should you clean the dishes right after using them or should you let them pile up before finally cleaning them would arise.

If you are the type who does not want seeing dishes piling up in the sink, then you will most likely clean them up after each use. In case you don't own a dishwasher, then the best course of action would probably be letting the dishes pile up for a while before cleaning them. The main reason is that cleaning the dishes, just like other manufacturing procedures, come with reducing costs to scale.

You may not probably realize it, but this chore actually comes with a fixed setup cost. Such cost comprises of the need to put water in your kitchen sink and roll up your sleeves every time you do the task. This means that you can do better if you are able to reduce the weekly fixed costs.

However, the above argument or scenario will break down if you do not put water in the dirty dishes immediately. Note that it would be harder to remove dried foods in the dishes. With this fact in mind, you can start comparing the effect of increased fixed cost against the increased effort and time needed to clean up every dish.

The best course of action in this case is to add water in the piled up dishes, in case you intend to clean them up later, to prevent the remaining foods from drying. This can definitely help you save up on the fixed costs without increasing the time and effort that you need to exert for such task. Apply this logic in other household chores, and you will surely have a better experience accomplishing them.

In Auctions

Are you planning to auction off one of your assets, for example an art piece? If you answer yes to this question, then it is advisable to decide carefully on the kind of auction that you will use. Note that there are currently numerous ways to set up an auction. However, there are those who recommend utilizing the second price auction if maximizing returns is the main goal. This type of auction allows every participant to bid on a sealed number. The one who placed the highest bid wins.

Note, however, that the one who offered the highest bid is not required to pay the highest bid price. What he will pay is the amount of the second highest bid. This option serves as an incentive for auction participants since they can bid higher, while still having the chance to lower their actual payment in case they win.

This can also benefit you because it optimizes your possible returns by forcing the auction participants to bid higher than what they normally do in similar situations. This makes game theory a truly useful technique in such particular case.

Negotiating in the real estate industry

If you need to negotiate in the real estate market, then game theory can also help you a lot. For instance, if the realtor you are dealing with says that you are currently in a multi-offer situation right after submitting your bid, then note that you have three options to deal with it. These options including sticking to your original bid, withdrawing your offer and offering a higher bid.

Because the industry will most likely choose a winner based on who offered the highest bid, the most probable course of action in order to win will be the third one. With this in mind, bid after calculating the reasonable amount it takes to get the property. Stick to that calculated bid. In case someone still beats you by bidding higher, then avoid beating your opponent.

Note that you have already done your best and calculated the odds to ensure that you will make an optimal play. Losing in this case will not cause any dissatisfaction on your part because you know that you did everything correctly. Instead of being upset from losing the bid, thank yourself and the principles of game theory from stopping you to listen to your impulse and formulate incorrect decisions that you will most likely regret in the end.

Chapter 9: Improving Relationships with Game Theory

Your knowledge about the game theory can also help a lot in improving your relationships. Marriage relationships, for example, can greatly improve with the application of game theory principles. If you have a somewhat chaotic relationship, one wherein you and your partner are constantly in a standoff with no one backing down, then the game theory may just be the solution that you are looking for.

Game theory and romantic relationships actually have a lot in common. Both concepts require the involvement of more than one person. Both also involve people who try their hardest to further their own gains to no avail, since the presence of the other party pose limitations to what they can do. Both concepts are also similar in the sense that there is a chance to develop a cooperative strategy wherein the parties involved are willing to work together in order to find a reasonable solution to the problem.

However, both also have the risk of implementing the non-cooperative strategy wherein the parties involved are unwilling to back down. In the two concepts, the non-cooperative strategy tends to be the most tempting, but it has grave results and consequences including the possibility of death. The cooperative strategy, on the other hand, may annoy the parties in both situations, but the results are always rarely fatal.

If you are in a romantic relationship, then questions like whether you should cooperate with your partner or not in case problems start to arise may come across your mind. This is when you can start applying the principles of game theory. One advantage of this is that the technique can tackle those situations wherein it would be impossible to have everything you want, but you still like to try reaching the best outcome possible.

One thing that the game theory teaches those who are in a relationship is that the relationship that they have is not all about working so hard to get everything that they want. It is actually having all that you possibly can under the current circumstances. In a romantic relationship or marriage, the current circumstance is quite obvious, but often overlooked, and that is the reality that another party is involved - your partner or spouse who also happens to want to get the best possible outcome from the situation.

If you are constantly in conflict with your partner, then the following game theory strategies can help you in overcoming the problem:

Think ahead

This means that you have to start thinking ahead by considering the possible consequences of an action that you plan to take. For instance, reflect on how your partner will most likely react on something that you are about to say. You can then think about how the reaction of your partner will influence your behavior now and in the future.

Think about the past

This means that you also have to look back and learn from the consequences of the actions that you made in the past. For example, you can go back to the time wherein you told your partner about the thing that you are planning to tell him now. What was his or her reaction during that time? If his or her reaction in the past is unfavorable, then ask yourself if there is something that you can do to prevent the same reaction or outcome.

Empathize

This means that you should start putting yourself in your partner's shoes. Consider what your reaction would be in case the situation is reversed. However, avoid reflecting on the situation based on who you are. Note that you and your partner have different personalities, so there is a great possibility that you will also react differently on the situation. Consider your partner's personality so you can better empathize with him or her.

The mentioned strategies are all reasonable. However, there are instances when you will most likely do the opposite due to the heat of the scenario. This case would trigger the implementation of the non-cooperative strategy. Try to avoid this as much as possible. Note that it is possible to reach and implement a cooperative strategy that will lead to romantic dinners, better relationship and other great things that couples can enjoy if you and your partner just put your heart and mind to accomplishing such goal.

A wise tip to make it easier for both you and your partner to implement the mentioned game theory strategy is to try to change the rules of the game. You can do so by devising incentives that will motivate the two of you to cooperate, instead of working hard to win the argument. For instance, if your arguments often come from being unable to fulfill household chores, then the best course of action could be to have a specific schedule for a particular task.

Impose a penalty in case one of you was unable to fulfill his or her designated duties for a particular day. Choose to impose a penalty that is more unpleasant than the task that your partner is dreading to do. This way, the two of you will be motivated to meet your scheduled responsibilities.

You can apply similar techniques in other areas of your life wherein you need to establish a stronger connection to people. This means that you can also implement the game theory to improve your relationship with your friends and family.

Chapter 10: Game Theory in the Office

Game theory is also applicable in the workplace. You can use it to negotiate for a raise or to thrive in your career. Since the game theory allows you to act, interact and decide in strategic settings, you have the chance to use it to get a good and rewarding job. This chapter will cover a few techniques on how you can implement it at work.

Perfect time to negotiate for a raise

If you have been working for a specific job for quite some time and you noticed that your salary remains the same, then it is greatly possible that you have already asked yourself when is the perfect time for you to approach your boss and negotiate for a raise. One thing that you should take note of when planning to use game theory when requesting for a salary increase is that you will encounter a number of mathematical models that you can use for bargaining.

Two vital factors can also identify who among the parties involved can get the better output from the bargain. The first factor you have to consider is that the party with less to lose is often the one who receives a more favorable deal. For instance, if there are more than five or ten workers in your office with skills that are similar or better than yours, then your boss will receive only a small incentive when he agrees on your request for a raise.

However, if there is another job waiting for you that is as good or better than your current one, then you can go to your boss and ask for a raise without hesitation. This is so since you know that will have a fallback in case he denies your request and you decide to quit. The second vital factor or feature is that the one who has more patience will most likely receive a bigger share.

If you are planning to use game theory to get a raise, then note that the perfect time to do so is when the two mentioned features or factors are in favor of your current situation. This means that you should only go to your boss and demand for a raise if he has more to lose when you quit and if you know that you have more patience than him.

Instead of waiting for the time when you will most likely get the raise that you want quickly, ask only when you know that you have more patience to drag the negotiation out.

Real-world Negotiations

If your current position in the workplace requires you to negotiate with other people on a regular basis, then it pays to master the principles of game theory since these can help you in this area. You can actually derive a few useful principles from game theory that can maximize the results when performing real-world negotiations. Some of these are the following:

Minimize Risk

In every situation that requires you to negotiate, it pays to minimize your risk by trying to reduce the maximum payout that the other players or your opponents can receive. You have to do so even if it requires assuring the opponent of a higher minimum.

Offer him something first

Making the first offer goes a long way in real-world negotiations. It is more beneficial to offer something first unless your current situation shows that you have a demonstrable informational disadvantage.

Establish trust

You have to improve your ability to commit like making promises and threats. You have to know exactly how to establish trust if you want to win a negotiation. Note that your capacity to win will most likely diminish if you cannot show the other players how credible you are. There are even instances when incurring expenses is better than losing your credibility.

Be consistent

You have higher chances of winning negotiations through game theory if you implement open-ended commitments. Note that this type of commitment encourages reliability. Short-term business relationships, on the other hand, breed duplicity. Be consistent when dealing with the people you are negotiating with. Instead of cutting ties right after winning or losing a negotiation, try to continue building a long-term relationship. This is beneficial since you never know when you will need their help again.

Learn the importance of reciprocity

The best way to describe reciprocity is that what you get is the result of what you give. This means that if you want to earn the respect of the people you are negotiating with, then you should also show them how you respect them. This way, you will have an easier time gaining their trust.

Midnight Dilemma

Another game theory technique that you can apply in the office is the midnight dilemma, but before delving into this technique, it is wise to understand how the prisoner's dilemma, the most popular game in the field, works. In prisoner's dilemma, you can imagine being one of the two criminals who are currently under police investigation. The most likely scenarios would be the two of you will remain silent, one of you will cooperate or the two of you will cooperate.

Depending on your actions, the payoff would either be a major or minor sentence. If you want the payoff to work solely in your favor, then there is a possibility that you will end up defending yourself and putting the blame on the other party. While you will still have some jail time when you cooperate that way, you can expect it to be lower than if you decided to keep silent.

You can also use the same tactic in various situations in the workplace. For example, all analysts in an investment company or bank can get a better payoff individually if everyone agrees to avoid staying past midnight. However, if the company starts offering an incentive for the employees' hard work, then each analyst will be better off staying late in the office and working overtime.

If, for instance, you decide to leave the office on time, instead of working overtime, then the management may label your performance as poor or bad. Since you know the consequences, the most likely result is that you will also work for long hours. This is a good thing especially if you are after the incentive that the management has decided to offer to their hardworking analysts.

Avoid snitching

This is another principle that can help you make the game theory work in your favor. Note that many times, the principles of game theory show that it is better to cooperate if you want to get positive results that will work for the long-term. If you repeatedly play the prisoner's dilemma, meaning you tend to defect at each opportunity you get, you will end up alone eventually, with no one in the office willing to partner or work with you.

You may gain instant rewards for snitching on your colleagues who are slacking off (this means that you repeatedly tell the authorized personnel in the workplace who are not performing well), but this will eventually backfire. Note that you may turn into an outcast in the end. Know when is the best time to take action and when is the best time to mind your own business.

If possible, leave the job of checking out other's performance to the right people. If you are not part of the human resources department whose tasks include evaluating the performance of the employees, then it is better for you to just mind your own business and do the tasks designated for you. This will prevent you from losing the trust of your colleagues.

Let your imagination work

There are even instances when acting crazy will pay off when you are in the workplace, especially if the output for doing such an act is favorable for the entire company. In the workplace, consistency is valued and considered as an asset. Most people love to work with reliable colleagues. However, when dealing with strategic settings or negotiations, your unpredictability will pay.

This principle is applicable to small-scale standoffs. An example would be pay negotiations. Sometimes, it is beneficial if your employer thinks that you would get mad and instantly leave your job if he turns down your request. This is so even if leaving your job is not on top of the things that you plan to do at the moment. In fact, it does not matter whether you have real intentions to quit the job.

What matters in this scenario is that your boss perceives the threat as realistic. This makes him factor in that possibility when calculating or deciding whether to approve your demand for a raise. Sometimes, it also pays to build a reputation as someone who is a ruthless risk-taker. This can greatly contribute in increasing the weight of negotiating threats.

Conclusion

Thank you again for downloading this book!

I hope this book was able to help you learn everything that you have to learn about the game theory. Game theory existed for quite a long time already and it has helped a lot of people achieve success in various aspects of their life.

The next step is to put the game theory principles and knowledge that you gained from this book into practice. Note that there are numerous situations wherein you will find the theory useful. With the tips, tricks and concepts that you have learned from this book, you can easily make the most out of applying game theory into your life.

You will have an easier time using it to make strategic decisions that will surely have a positive impact in your life, your future, and the way you interact with the people around you.

Finally, if you enjoyed this book, please take the time to share your thoughts and post a review on Amazon. It'd be greatly appreciated!

Thank you and good luck!